SACR
Sacramento, CA 95814
04/20

This Book Was Purchased With
Funds Generated By
The Library's Passport Center

SACRAMENTO PUBLIC LIBRARY

GETTING THE JOB DONE

EXTERMINATORS

Therese M. Shea

PowerKiDS press

New York

Published in 2020 by The Rosen Publishing Group, Inc.
29 East 21st Street, New York, NY 10010

Copyright © 2020 by The Rosen Publishing Group, Inc.

All rights reserved. No part of this book may be reproduced in any form without permission in writing from the publisher, except by a reviewer.

First Edition

Editor: Elizabeth Krajnik
Book Design: Reann Nye

Photo Credits: Cover, pp. 13, 21, 22 hedgehog94/Shutterstock.com; p. 4 luis2499/Shutterstock.com; p. 5 torook/Shutterstock.com; p. 7 Justin Sullivan/Getty Images News/Getty Images; p. 8 Odua Images/Shutterstock.com; p. 9 Simon Battensby/Photographer's Choice/Getty Images; p. 11 Chuck Wagner/Shutterstock.com; p. 12 Adriano Kirihara/Shutterstock.com; p. 14 Andrey_Popov/Shutterstock.com; p. 15 NATTAPON JUIJAIYEN/Shutterstock.com; p. 17 fstop123/E+/Getty Images; p. 18 Dan Olsen/Shutterstock.com; p. 19 bruceman/E+/Getty Images.

Library of Congress Cataloging-in-Publication Data

Names: Shea, Therese, author.
Title: Exterminators / Therese M. Shea.
Description: New York : PowerKids Press, [2020] | Series: Getting the job done | Includes index.
Identifiers: LCCN 2018053285| ISBN 9781725300002 (paperback) | ISBN 9781725300026 (library bound) | ISBN 9781725300019 (6 pack)
Subjects: LCSH: Pests–Control–Vocational guidance–Juvenile literature. | Household pests–Control–Vocational guidance–Juvenile literature.
Classification: LCC TX325 .S54 2020 | DDC 628.9/6023-dc23
LC record available at https://lccn.loc.gov/2018053285

Manufactured in the United States of America

CPSIA Compliance Information: Batch #CSPK19. For Further Information contact Rosen Publishing, New York, New York at 1-800-237-9932.

CONTENTS

A QUICK CALL AWAY................ 4
ON THE JOB...................... 6
PEST PROFESSIONALS 8
THE TOOLS FOR THE JOB 10
PROTECTIVE GEAR 12
ALWAYS DIFFERENT 14
EXTERMINATOR EDUCATION 16
EXAMINING THE NUMBERS 18
UP TO THE CHALLENGE?............22
GLOSSARY........................23
INDEX 24
WEBSITES 24

A QUICK CALL AWAY

Imagine coming home after a week of vacation. You step through the door and … you see mice running everywhere! Gross! You need to call an exterminator—and fast. Luckily, exterminators are ready to help people get rid of every kind of pest you can think of, from creepy cockroaches to revolting rats.

Fascinating Career Facts

Exterminators can be lifesavers because some pests aren't just annoying—they spread diseases, or illnesses, to people.

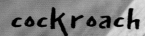
cockroach

You can get rid of one or two mice with a trap. More than that, and you may need to call a **professional** exterminator.

In this book, you'll learn that exterminators, also called pest control workers, do more than exterminate, or remove, pests. You'll discover how they tackle different kinds of jobs and what they do to prepare for this career. Do you have what it takes to be an exterminator? You're about to find out!

ON THE JOB

On a normal workday, an exterminator gets a call from a **client** at a home or business asking for help. The van or truck is packed with the necessary **equipment**. The exterminators **inspect** the property—inside and outside—to find the pests and discover how much damage, or harm, they've done. Then they'll figure out the best way to tackle the problem and talk to the client about any choices they have.

Once the method of pest removal is chosen, pest control workers carry it out. Then, they'll clean up the site and offer advice about how to stop the pests from returning.

> This exterminator is using a specially trained dog to find bedbugs!

PEST PROFESSIONALS

Exterminators who **specialize** in certain tasks may have special job titles. Exterminators called pest control technicians have several duties. The technicians find out if a place has a pest problem, locate where the pests are, and figure out the best way to get rid of them. They also suggest ways to stop pests from reentering the area.

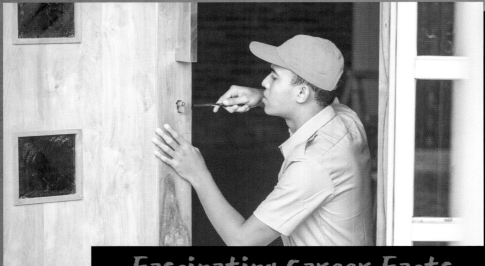

Fascinating Career Facts

Pest control workers may also do a bit of construction! They may repair buildings that have been harmed by pests such as termites or construct something to keep out pests such as rats.

Sometimes the pests are in one place, such as these wasps that built a huge nest in this house! Other times, the pests are spread out.

While pest control technicians may use some weaker **pesticides**, pest control applicators are the exterminators who use the strongest chemicals. Applicators are specially trained to use powerful pesticides in a way that won't harm people. They may use poisoned bait or a gas called a fumigant.

THE TOOLS FOR THE JOB

Pest control applicators that use fumigants are called fumigators. They follow a process before releasing, or letting free, the fumigant. First, people, animals, and plants are removed from a building. Then, places where gas could escape—such as windows, doors, and vents—are sealed off with plastic or tape. A special kind of tent may be placed over the entire building. Finally, the fumigators use a hose to pump the building full of the poison gas.

Fascinating Career Facts

You may think that pest problems are a recent issue. Actually, pests have been a problem for as long as people have been on Earth! As early as 2500 BC, people used mixtures, oils, and plants to kill bugs.

Fumigants are sometimes the only way to get rid of pests such as certain bugs that live within walls.

The fumigant is specially made so that it will remove certain kinds of pests and then break down. That way, no harmful chemicals are left that could make people sick.

PROTECTIVE GEAR

Pesticides are generally the last **resort** when it comes to fighting pests. The U.S. Environmental Protection Agency (EPA) suggests first cutting off pests' food and water sources and their entryway into buildings. However, that doesn't always work or isn't always possible. That's when pesticides are used.

After the pesticide is released, it's important for the exterminator to carefully wash themselves and their gear. Some protective suits are thrown out rather than cleaned.

When pest control professionals work with pesticides, they must wear the correct clothes and use the right equipment to protect themselves. Chemicals can enter the body by being touched, breathed in, or even through the eyes. For the most toxic pesticides, the exterminator wears a body suit with a hood, gloves, goggles, shoe coverings or boots, and a **respirator**.

ALWAYS DIFFERENT

There's no such thing as a regular day as an exterminator! If you want to make sure you're not bored at your job, being an exterminator might be for you. Every job is different. Every location is different. Every pest problem is different, too.

Fascinating Career Facts

Exterminators learn about the life cycles of pests. They know where they might make nests, lay eggs, and other behaviors. They know the best places to put traps and apply pesticides.

Exterminators don't always kill pests. Live traps capture rats and mice so they can be released, unharmed, somewhere else.

Mosquitoes and ticks may require chemical treatments, such as sprays. A termite problem may demand bait systems, sprays, or fumigation. Bedbugs are sometimes removed with heat or pesticides. Rats and mice may require traps with bait. While people could try these methods themselves, professionals know the most effective way of getting rid of pests and sometimes **guarantee** good results.

EXTERMINATOR EDUCATION

If you're interested in becoming a pest control worker, you'll need to be at least 18 years old and have a high school education. Most exterminators begin their career as a pest control technician. They train while they're on the job, under more **experienced** exterminators. They also take classes and learn how to safely use pesticides.

Pest control workers generally need to be **licensed**, and certain states may require them to take exams. If they specialize in using special methods of pest control, they may do even more training for special licenses. Exterminators may continue to take classes as new pest-control methods are invented.

As pest control workers gain more experience, they may supervise, or oversee, other exterminators.

EXAMINING THE NUMBERS

As of 2016, there were around 79,000 pest control workers in the United States, according to the U.S. Bureau of Labor Statistics. It's believed that this number will grow by 8 percent in the coming years, possibly to around 85,400 by 2026. These numbers tell us that pest control is a career with a future.

Fascinating Career Facts

The Formosan termite is one **invasive species** that homeowners fear. They're sometimes called "super-termites" because they eat so much wood. They live in colonies of millions!

Termites can cause a lot of damage to a house very quickly. People often hire professionals to deal with these pests.

One of the reasons for this growth is that the number of invasive species in the United States is rising, too. For example, Asian brown marmorated stink bugs didn't arrive in the country until 1996. These pests can now be found in most of the United States.

INCOME AND HOURS

In 2017, the average salary for pest control workers in the United States was $34,370, according to the U.S. Bureau of Labor Statistics. That means half of all exterminators earned more than this amount and half earned less. Like most jobs, the more experience a worker has, the more they'll likely earn.

Most pest control workers work full time, or at least 40 hours a week. About 20 percent of exterminators work more than this. They often work on weekends and at night because people who need exterminators often want the problem taken care of immediately! Working overtime or at odd hours may mean more money, however.

> Do you want to start your own pest control business someday? The business may make $50,000 to $75,000 in the first few years. However, you'll also need to spend money on costs such as a van and equipment.

UP TO THE CHALLENGE?

There are a few **qualities** that are important in all pest control professionals. They should be good data keepers, pay attention to details, and follow directions well. They also have to be physically fit, since they'll be on their feet a lot. They'll need to bend down to pest level—and even crawl into tight spaces to reach problem areas.

Exterminators also must have good communication skills. They must be able to get information from clients and explain their methods of pest control clearly. Satisfied clients will tell others about excellent service—and the jobs will keep rolling in!

GLOSSARY

client: A person who pays a professional person or organization for services.

equipment: Supplies or tools needed for a special purpose.

experienced: Having gained skill or knowledge by doing something.

guarantee: A promise that the quality of something will be as good as expected.

inspect: To look at something carefully in order to learn more about it.

invasive species: A kind of animal or plant that is not native to a place and does harm as it spreads.

licensed: Having official permission to have or do something.

pesticide: A chemical that is used to kill animals or insects.

professional: Someone who does a job that requires special training, education, or skill.

quality: A characteristic or feature that someone has.

resort: Something chosen for help.

respirator: A tool worn over the mouth and nose so that you can breathe when there is something harmful in the air.

specialize: To limit your business or area of study to one subject.

INDEX

B
bedbugs, 6, 15

C
chemicals, 9, 11, 13, 15
classes, 16
client, 6, 22

D
damage, 6, 19
diseases, 4

E
education, 16

F
fumigant, 9, 10, 11

H
hours, 20

M
money, 20

P
pest control technicians, 8, 9, 16
pesticides, 9, 12, 13, 14, 15, 16

Q
qualities, 22

S
suit, 13

T
termites, 8, 15, 18, 19
traps, 5, 14, 15

U
U.S. Bureau of Labor Statistics, 18, 20
U.S. Environmental Protection Agency (EPA), 12

WEBSITES

Due to the changing nature of Internet links, PowerKids Press has developed an online list of websites related to the subject of this book. This site is updated regularly. Please use this link to access the list: www.powerkidslinks.com/GTJD/exterminators